T0275965

SpringerBriefs in Energy

More information about this series at http://www.springer.com/series/8903

Bruno G. Pollet · Iain Staffell
Kerry-Ann Adamson

The Energy Landscape in the Republic of South Africa

 Springer

Bruno G. Pollet
Eau2Energy
Nottingham
UK

Kerry-Ann Adamson
4th Energy Wave Ltd.
Edinbugh
UK

Iain Staffell
Imperial College Business School
London
UK

ISSN 2191-5520   ISSN 2191-5539   (electronic)
SpringerBriefs in Energy
ISBN 978-3-319-25508-8   ISBN 978-3-319-25510-1   (eBook)
DOI 10.1007/978-3-319-25510-1

Library of Congress Control Number: 2015950924

Springer Cham Heidelberg New York Dordrecht London

Printed on acid-free paper

Springer International Publishing AG Switzerland is part of Springer Science+Business Media
(www.springer.com)

Preface

Energy is an important driver for economic growth, and the Republic of South Africa (RSA) finds itself gripped by a prolonged energy crisis. Chronic under-investment in the electricity sector has led to escalating power prices and a short-age of capacity during peak demand periods, leading to demand rationing and blackouts.

This brief provides an overview of the energy landscape in RSA, provid-ing background and context to the current situation, and presenting analysis of the policies being put forward by the government for the near future. Four broad areas are covered: reserves and production of fossil fuels, the electricity sector, the rapidly growing exploitation of renewable energy and the recent push towards developing an industry around hydrogen and fuel cells.

The following section of this brief presents a methodical discussion of the energy landscape in RSA, covering the general situation, the supply and demand for energy, and the structure of the energy sector (Chaps. 1 and 2). Chapter 3 presents data and analysis of the country's fossil fuels, electricity generation and potential future sources of production and industry. Chapter 4 discusses recent developments, including the impact on green jobs and green funds, and Chap. 5 reflects on the policies that have been proposed and their potential implications. Chapter 6 puts forward some concluding remarks and recommendations.

Contents

Abstract

The Republic of South Africa (RSA), often seen as the "Powerhouse of Africa" is facing major energy challenges. (Un)planned outages, energy shortages, black-outs, high energy tariffs, many years of underinvestment in power infrastructure and energy poverty in low-income households are the main issues. To eradicate this, the RSA government has rolled out several energy and energy efficiency programmes and initiatives with a strong emphasis on off-grid renewable energy solutions, hydrogen energy, and oil and gas (including shale) exploration opportunities. Since 2013, most of the country's top construction projects have focussed on renewable energy projects. The country's aspiration is to build up to an "industrial revolution" incorporating the development of a "Green Economy" which could significantly boost the nation manufacturing base. The aim of this brief is to provide a fact based analysis of the current energy landscape in RSA. This can then be used alongside the aspirations and grand challenges the country has identified to ascertain the scale of the issues that the country faces. This brief is not set out to "solve" the problems of today's energy in RSA, but to provide the evidence needed to think more systematically about them.

Keywords Republic of South Africa (RSA) · Energy · Renewables · Hydrogen · Natural gas · Nuclear

Chapter 1
Introduction

The Republic of South Africa, a *BRICS* middle-income developing country, faces many issues to economic growth such as: (i) energy challenges; (ii) aged and inadequate infrastructures; (iii) inefficient regulatory processes delaying inter/national and local investments; and (iv) inefficient government co-ordination, long-term planning and vision (which in turns contributes to investors' uncertainty—although in 2013 RSA received US$8 billion in Foreign Direct Investment). According to the World Bank, RSA's economy expanded to a mere 1.5 % in 2014, the slowest pace since the 2009 recession. In views of kick-starting the economic growth, the RSA government has recently developed a '9-Point Plan' comprising simultaneous actions in key strategic areas [1].

Energy is an important driver for economic growth, and RSA finds itself gripped by a prolonged energy crisis. Chronic under-investment in the electricity sector has led to escalating power prices and a shortage of capacity during peak demand periods, leading to demand rationing and blackouts. In response to this national energy crisis, in early 2015 the government created the so-called 'Energy War Room' to urgently and systematically implement the Cabinet's '5-Point Energy Plan', which consists of: (i) maintaining the country's state-owned electricity company *Eskom* (with a projected financial bailout of US$1.9 billion for FY2015/16); (ii) introducing new generation capacity through coal; (iii) partnering with the private sector into co-generation contracts; (iv) introducing gas-to-power technologies; and, (v) accelerating the demand side management.

This brief provides an overview of the energy landscape in RSA, providing background and context to the current situation, and presenting analysis of the policies being put forward by the government for the near future. Four broad areas are covered: reserves and production of fossil fuels, the electricity sector, the rapidly growing exploitation of renewable energy, and the recent push towards developing an industry around Hydrogen and Fuel Cells as well as Platinum Group Metals (PGM) beneficiation.

© The Author(s) 2016
B.G. Pollet et al., *The Energy Landscape in the Republic of South Africa*,
SpringerBriefs in Energy, DOI 10.1007/978-3-319-25510-1_1

The following section of this brief presents a methodical review of the energy landscape in RSA, covering the general situation, the supply and demand for energy, and the structure of the energy sector (Chaps. 1 and 2). Chapter 3 presents data and analysis of the country's fossil fuels, electricity generation, and potential future sources of production and industry. Chapter 4 discusses recent developments, including the impact on green jobs and green funds, and Chap. 5 reflects on the policies that have been proposed and their potential implications.

Reference

1. Department of Communications (2015) Economic sectors, employment & infrastructure development cluster post SoNA media briefing. http://www.gcis.gov.za/print/8394

Chapter 2
Energy and Africa

Abstract This chapter focuses on the current energy landscape in Africa as well as the challenges the continent is currently facing. It also highlights the energy demand and resources, the renewable energy policy, investment and corruption.

Keywords Energy · Africa · Renewables · Investment · Corruption

South Africa is only one out of 54 countries making up the African continent. Now with over 1.1 billion people, the continent is being seen by China, Europe and increasingly the US, as the next major trading partner. This in itself represents a tectonic shift of emphasis away from the more historic image of the continent as a passive recipient of development funds. Whilst the continent still faces any number of challenges, in terms of energy, transport and deployment of the critical Information and Communication Technology (ICT), all of which are somewhat inextricably linked, the focus on a value add for Africa itself is underway.

One of the biggest challenges policy makers, aid bodies and analysts in this area face is the lack of in-depth up-to-date information on what the on the ground energy issues are. With gathering of statistics being far down the chain of day-to-day needs, it is only in very recent times that any kind of systematic gathering of data on the energy landscape, needs, outputs and inputs from Africa has started to change. Of the data available the World Bank and the International Energy Agency (IEA) are the two best sources, but even they admit that there are a number of gaps, including an understanding of the actual scale of use of back-up power generators and the level of use of more traditional power sources, such as wood stoves and biomass. Along with the overall energy landscape in Africa this is changing, but often when using data on Africa there is a caveat of a 50 % statistical variance on the information, which could in itself have major ramifications for policy!

© The Author(s) 2016
B.G. Pollet et al., *The Energy Landscape in the Republic of South Africa*,
SpringerBriefs in Energy, DOI 10.1007/978-3-319-25510-1_2

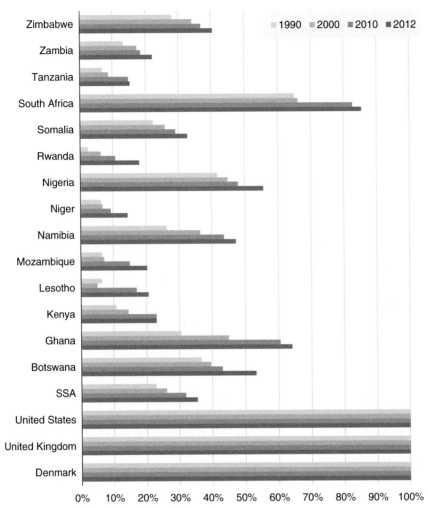

Fig. 2.1 Access to electricity for selected SSA countries, 1990–2012 [1]

As of today though, and mentioned earlier in the brief, it is thought that 65 % of the population has no access to electricity. This though is very unevenly distributed across the continent, with North African countries having over 90 % access to electricity, but in sub-Saharan Africa (SSA) in many countries this drops to 50 % or under. Figure 2.1, using World Bank data, shows the increase of access to electricity over time for selected countries in SSA. For reference Denmark, the US and UK are used as comparators across charts in this section.

As the waterfall chart clearly shows, RSA is in a comparatively strong position, especially when compared with a number of its trading neighbours in the SAPP. Using the World Bank data, up to 2012, Tanzania was the worst off in the Southern African Power Pool (SAPP, see later), with only 15 % of the country's population having access to electricity and Zimbabwe, Namibia and Botswana were all between 40 and 53 %.

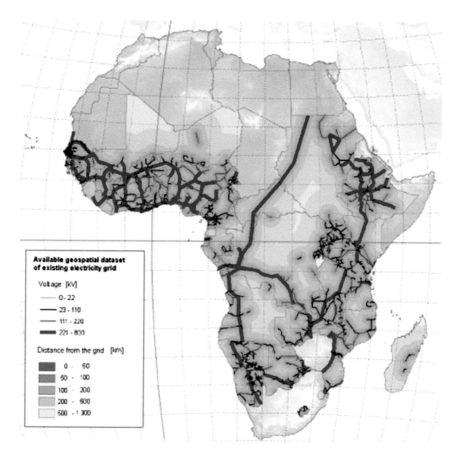

Fig. 2.2 Geospatial mapping of electricity power lines in SSA, 2011 [2]

To put this in content though, according to the IEA data over 145 million people have gained access to electricity since 2000. This has been led by improvements in Nigeria, Ethiopia, South Africa, Ghana, Cameroon and Mozambique.

One of the clear reasons for the lack of access to power has been till date the fundamental lack of a modern energy infrastructure. Figure 2.2, a much quoted chart by development agencies, clearly shows that whilst some trunk networks do exist, the connecting infrastructure for residential use, at between 2 and 30 kV, is non-existent in the many parts of the continent.

For settlements that have no gird access the more traditional energy use is that of biomass, or biofuels. As quantifying the level of this use is agreed up by the development and statistical agencies as being on the largest challenges in base-lining the current energy mix in Africa, the data shown here need to be taken with some caution. The key point to take from this chart is the level of reliability that many countries in Africa still have on traditional forms of energy as part of their Total Primary Energy Supply (TPES). Figure 2.3 shows the biofuels and waste as part of selected African countries' Total Primary Energy Supply.

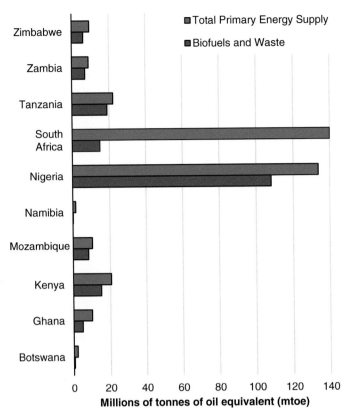

Fig. 2.3 Biofuels and waste as part of selected African countries' total primary energy supply, 2012 [3]

Within this mix, it can once again be very clearly seen the difference in development between RSA and other African countries. What is perhaps surprising is the level of demand for traditional energy sources in the fossil fuel rich nation of Nigeria. The country, which is a net energy exporter of crude oil, has a residential sector which consumes over 95 % of the total of biofuels and waste per annum (Fig. 2.3). This highlights in one sentence the challenge faced by a number of African nations. The difference is between the value to the economy of selling, and exporting its mineral and energy resources, or using them at home.

In terms of use of renewable energy, adoption of non-hydro traditional renewables has so far been very low in Africa, in general. Whilst on a global level the move is a transition to a renewable-based low carbon economy, the cost and prohibitive lack of infrastructure in many parts of Africa has seen adoption at a very low level.

As with many low carbon technologies cost really is an issue, and although when compared with say the price of diesel, many economic cases can be made, the on the ground reality is that till date distributed energy, and micro grids remain so niche that reliable statistics on adoption, and deployment data, are thin to zero on the ground. According to one IEA dataset [3], adoption of solar PV in the whole of Africa in 2012 was below the threshold to be broken out into its own category and is published together with the data from the Middle East at just 300 installations.

2.1 Energy Demand and Resources

According to the International Renewable Energy Agency (IRENA) [4] just five countries in Africa dominate the current power market (Fig. 2.4). These are South Africa at 21 % of primary energy use, Nigeria 16 %, Egypt 11 %, Algeria 6 % and Ethiopia 5 %. It should be noted that this report uses data from 2009. The rest of the African continent represents only 41 % of total primary energy use.

When we unpack the data from 2012 for these countries we can see that, reading bottom to top, in terms of energy consumption till date, residential and transport energy consumption makes the majority of demand in the majority of African nations. This is primarily due to the lack of high energy consuming industry, aside from mining.

Figure 2.5, again mining data from the IEA, clearly shows this dominance of residential and transport sectors. Note that Nigeria's residential sector has been removed from the dataset, as at 88,935 ktoe (kilotonnes of oil equivalent) it dwarfs all other sectors.

When these are compared with 24,906 ktoe from industry, 16,690 ktoe from transport and 16,500 ktoe from the residential sector in South Africa [3], the scale

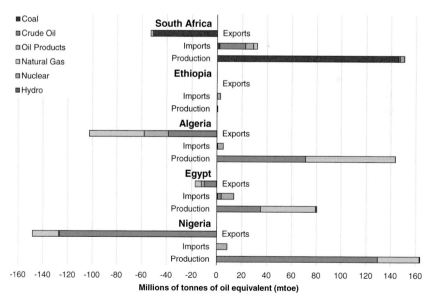

Fig. 2.4 Energy production, import and export from South Africa, Nigeria, Egypt, Algeria and Ethiopia, 2012 [3]

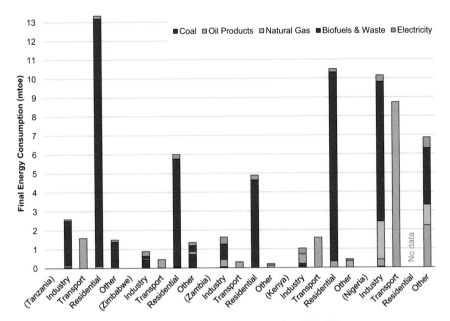

Fig. 2.5 Energy demand by end use, selected African countries, 2012 [2]

of the difference between the energy equation in South Africa and most of the rest of Africa becomes clear.

Africa is not an energy poor continent. Whilst it is not as rich as say the Middle East, Nigeria, Libya, Angola and Algeria especially are locally oil rich. For coal South Africa is the only key supplier across the whole of the continent and natural gas, so far at least, is not a player (Table 2.1).

With boom times in North America over shale oil and gas it is perhaps understandable that there has been a flurry of activity in ascertaining if African holds any potential in this area. According to the EIA, as reported by KPMG [6] that:

- Algeria has 707 trillion cubic feet (almost 20 trillion m^3) of 'technically recoverable shale gas resources';
- Libya has estimated reserves of 122 trillion cubic feet of shale gas, the the 5th largest deposit of shale oil in the world;
- Egypt has reserves of 100 trillion cubic feet;
- Tunisia, Morocco and Western Sahara also all have proven reserves and
- South Africa has 390 trillion cubic feet of shale oil.

The question of economic viability to extract and process is very much at early stages with many countries starting a programme of drilling and testing.

Turning to renewables, IRENA raises the issue of technical, not economic potential, of renewable energy in Africa [5]. Table 2.2 reproduces an overview of the results. One note of caution that needs to be raised here is that IRENA states a 50 % uncertainty factor in the data.

Table 2.1 African energy proven reserves of oil, natural gas and coal, 2014 [5]

	Oil (2013)—thousand million barrels	Natural gas (2013) tcm	Coal (2013)—million tonnes
Algeria	12	5	
Angola	13		
Chad	2		
Rep. of Congo	2		
Egypt	4	2	
Equatorial Guinea	2		
Gabon	2		
Libya	48	2	
Nigeria	37	5	
South Sudan	4		
Sudan	2		
Tunisia	0		
South Africa			30,156
Zimbabwe			502
Other Africa	4	1	1156
Total Africa	*130*	*14*	*31,814*
Rest of World	*1571*	*172*	*859,717*

Table 2.2 Technical potential for power generation from renewables in Africa [5]

	CSP	PV	Wind	Hydro	Biomass	Geothermal
Central Africa	299	616	120	1057	1572	
Eastern Africa	1758	2195	1443	578	642	88
Northern Africa	935	1090	1014	78	257	
Southern Africa	1500	1628	852	26	96	
Western Africa	227	1038	394	105	64	
Total Africa	4719	6567	3823	1844	2631	88

As stated earlier, whilst there is significant technical potential for power from the sun, in both CSP and PV, in Africa the on the ground reality is that till date it remains a market with potential only.

2.1.1 Renewable Energy Policy

One of the key determinants for overseas investment in Africa, in the energy sector, is clear and actionable government policy. As with other countries in the world the signal from the government that it aims to modernise, liberalise and update the energy infrastructure in its country is a critical first step to moving forward.

In terms of energy and Africa there are few countries in Africa without a policy document looking at energy but in reality fewer of them are being acted upon.

Some relevant examples, with somewhat elastic timeframes are:

- The Nigerian government has put a legislative framework in place under the Renewable Energy Masterplan. Within this off-grid and distributed solar are actively being encouraged. According to some of the frameworks there will be a solar PV target of 500 MW by 2025.
- The Ghanaian government has a 10 % renewable energy target by 2020.
- Zambia has a rural electrification scheme, with a focus on deploying renewable energy. Within this the government is offering a range of fiscal incentives.

This is one area that would significantly benefit from future collaborative endeavours—cataloguing and bringing together policy makers from the different power pools to develop cross border, economic and effective policy on renewable energy, and energy trading in general.

2.1.2 Investment and Corruption

Finally, in this section on Africa the issues of overseas investment and corruption should be addressed. Corruption, which is endemic in some parts of Africa, is often quoted as one of the key roadblocks to unlocking increased overseas investment, in African power and transport projects. Transparency international each year releases a report indexing countries on their corruption levels. Within the index 100 is fully non-corrupt and 0 is fully corrupt. The results for a number of the African nations are shown in Fig. 2.6.

In the report Somalia was ranked the most corrupt country in the world.

In terms of investment the Infrastructure Consortium for Africa (ICA) has been collating and publishing data on investment in Africa in transport, water, energy

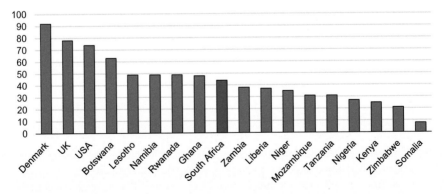

Fig. 2.6 Transparency international corruption scores for selected African countries, 2014 [7]

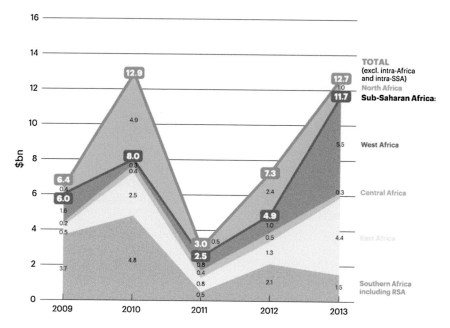

Fig. 2.7 Trends in energy infrastructure finance from ICA Members, 2009–2013 [8]

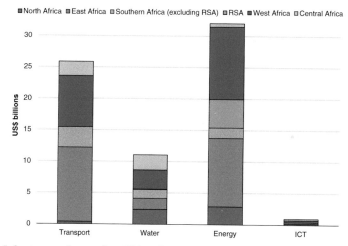

Fig. 2.8 Infrastructure finance from ICA and Non-ICA Members, 2013 [8]

and ICT sectors for the past 2 years. The ICA is a group of the following: the African Development Bank and Development Bank of Southern Africa, World Bank, International Finance Corporation (IFC), European Commission (EC) and European Investment Bank (EIB), the G8 (Canada, France, Germany, Italy, Japan, Russia, UK and the US), and the Republic of South Africa. They represent billions of annual investment in Africa.

The ICA secretariat published the chart given in Fig. 2.7 in the 2014 Annual Report. What the figure highlights is the continued cyclical nature of investment in energy. It should be noted that the 2013 data contains the US 2013 initiative "Power Africa". This represents an additional $7 billion from federal government for financial support and guarantees through its agencies, and has leveraged about $20 billion of private sector commitments.

As well as the ICA, China is one of the biggest investors in infrastructure projects in Africa, but it should be noted that over 50 % of the funding from China goes to transport projects, not energy. When the data from the ICA are combined with other non-affiliated countries and banks, we see that in 2013 energy infrastructure in Africa received $32 billion from overseas lenders (Fig. 2.8).

What is concerning in this chart is the lack of finance to central Africa, a region where energy poverty is at its worst.

References

1. The World Bank (2015) Statistics. http://data.worldbank.org/
2. Szabó S et al (2011) Energy solutions in rural Africa: mapping electrification costs of distributed solar and diesel generation versus grid extension. 2011 Environ Res Lett 6:034002
3. The International Energy Agency. Statistics. http://www.iea.org/statistics/
4. IRENA (2015) Prospects for the African power sector. https://www.irena.org/DocumentDownloads/Publications/Prospects_for_the_African_PowerSector.pdf
5. BP Statistical Review of World Energy (2015). http://www.bp.com/en/global/corporate/about-bp/energy-economics/statistical-review-of-world-energy.html
6. KPMG (2015) Oil and gas in Africa: reserves, potential and prospects of Africa. https://www.kpmg.com/Africa/en/IssuesAndInsights/Articles-Publications/General-Industries-Publications/Documents/Oil%20and%20Gas%20in%20Africa%202014.pdf
7. Transparency International, Corruption Perceptions Index (2014) Results. http://www.transparency.org/cpi2014/results
8. Infrastructure Consortium for Africa (2014) Annual report. http://www.icafrica.org/en/topics-programmes/energy/energy-investment-picture-data/

Chapter 3
The Current Situation in the Republic of South Africa

Abstract This Chapter focusses on the current energy landscape in South Africa as well as the challenges that the country is currently facing. It also highlights on the energy demand and resources.

Keywords Energy · South Africa · Renewables · Investment

Africa is home to six of the ten fastest growing economies, driven largely by foreign investment (inc. AfDB—African Development Bank) reaching ~US$150 billion in 2015. Sub-Saharan Africa's GDP is currently estimated at ~US$1.6 trillion (~US$2.3 trillion for Africa), similar to that of Spain, Australia or Texas (USA) [1]. However Africa is facing a major challenge: 1.1 billion people live on the continent (14 % of the world population) with 65 % of the population having no access to electricity. Sadly, Africa is still also home to the world's largest concentration of an impoverished population, and the considerable gap between "rich" and "poor" continues to widen. By 2050 Africa's population is estimated to reach 1.9 billion and consequently, energy supply and access will continue to be major issues [2, 3]. In order to meet the projected growth in electricity demand, Africa needs to add ~250 GW of new capacity between now and 2030, totalling to an annual capital cost of ~US$20 billion by 2030. This takes into account that during the next few years, Sub-Saharan Africa (SSA) (current population of ~960 million [1]) is expected to be one of the fastest growing regions in the world [2, 3].

RSA's GDP is ~US$350 billion, similar to that of Denmark or Malaysia [2]. In RSA, the percentage of households with access to electricity had increased by 8.2 percentage points between 2002 and 2012 to 85.3 %; however, a large number of households were still without electricity or could not afford to use adequate electricity to satisfy their needs [4]. According to the General Household Survey (GHS) in 2012 [5], 1.45 million (11 %) of RSA households did not have access to electricity, while another 0.6 million (3.6 %) households accessed electricity informally or illegally. Out of the 3.6 % without formal access to electricity, 73.1 % were connected to an informal source that the household paid for (e.g. sharing a connection with a neighbour), while 11.7 % made use of illegal connections [5].

It is estimated that the annual electricity demand in RSA will grow from 345 to 416 TWh by 2030, as opposed to the 454 TWh expected in the Integrated Resource Plan for Electricity document by the Department of Energy (DoE) [6]. According to Eberhard [7], this conservative demand assumption is "aspirational", having been aligned to the National Development Plan's expectation of an average annual growth of 5.4 % (see later) [8]. Moreover, compared with BRIC countries (Brazil Russia India China), RSA's economy is growing far slower, while electricity demand has receded back to 2006 levels. This is despite RSA's electrification rate (85.3 %) being higher than the world average of 80.5 %, the developing world (74.7 %) and SSA in particular (30.5 %) [9].

RSA is going through a rapid period of change and growth and plans to spend US$50 billion on clean energy in the coming years in an attempt to decrease reliance on coal-fired power plants, which provide 85 % of its electricity with the highest levels of CO_2 emissions [10, 11]. Since 2009, the RSA government has made a bold and challenging commitment to reduce the country's emissions by 34 % from *business-as-usual* levels by 2020 and 42 % by 2025 [12]. According to a fairly recent report by KPMG [13], RSA faces a 'carbon chasm'.

The power sector contributes ~45 % of the RSA's carbon emissions with an effective emissions cap set at ~275 Mt/annum CO_2 equivalent in the DoE's Integrated Resource Plan (IRP) [14]. The *post*-Apartheid RSA Government no

Table 3.1 Republic of South Africa key energy and consumption indicators: 2008–2011. *Source* Statistics South Africa 2014 and South African Department of Energy

Indicator name	Units	2010	2011	2012	2013
Population	Millions	50.1	51.8	51.6 (*est.*)	53.0 (*est.*)
Access to electricity	% of Population	75.8	84.7		84.0
CO_2 emissions from electricity and heat production	Million metric tonnes	473.6	461.6		
Electricity production	TWh	238.3	240.5	234.2	233.1
Electricity imported	TWh	12.2	11.9	10.0	9.4
Electricity exported	TWh	14.7	15.0	15.0	13.9
Coal consumption	TWh	1064	1028	1029	1026
Natural gas consumption	TWh	40.7	40.7	41.9	40.7
Hydroelectric consumption	TWh	2.1	2.1	1.2	1.1
Nuclear power consumption	TWh	12.7	14.2	12.6	13.9
Renewable energy consumption	TWh	0.3	0.3	0.4	0.3
Total energy consumption	TWh	1119.3	1085.3	1085.1	1082
Mining energy consumption[a]	TWh	13.2	13.9	13.4	13.7

[a]Anglo Platinum, Lonmin and Impala Platinum

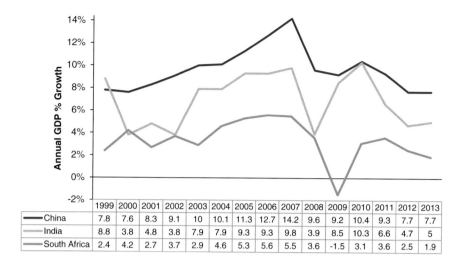

Fig. 3.1 RSA, India and China GDP growth rates. *Source* World Bank [9]

longer leaves power planning exclusively to the stated-owned utility company *Eskom* and the RSA DoE has been mandated to produce an electricity plan (the most recent of which is the IRP2010-30, see later) [15]. Not only has the current ANC (African National Congress) government promised access to electricity for all, it is determined to move the country out of low value-added, low skilled jobs to a knowledge-based and innovation economy. Together with *Eskom*, the DoE has embarked on a massive programme to bring the electricity supply and distribution system into balance. However this comes at a cost of an infrastructural price tag of around US$30 billion for the construction of new power stations, including Medupi in Limpopo and Kusile in Mpumalanga. After many years of barring *Eskom* from investing in new capacity, recently the RSA government approved a support package for *Eskom*, which will see the utility company raising over US$4 billion in additional debt and receiving an equity injection from the State. In this way it clearly wants to mirror the spectacular growth and change we have seen in China and India [16, 17].

The lack of access to guaranteed power has been identified by the RSA government as one of the key barriers to economic growth, and as such is under intense scrutiny. The country suffers from at times rolling blackouts, with the residential sector bearing the brunt to ensure power to heavy industry users continue. Table 3.1 and Fig. 3.1 provide an overview of RSA's energy sector and wider economy.

References

1. http://data.worldbank.org/region/sub-saharan-africa
2. Eberhard A, Rosnes O, Shkaratan M, Vennemo H (2011) Africa's power infrastructure investment, integration, efficiency. http://www.gsb.uct.ac.za/files/AfricasPowerInfrastructure.pdf
3. Eberhard A, Foster V, Briceño-Garmendia C, Ouedraogo F, Camos D, Shkaratan M (2008) Underpowered: the state of the power sector in sub-Saharan Africa, Background Paper 6, Africa Infrastructure Country Diagnostic, The World Bank, Washington, DC. http://www.infrastructureafrica.org/system/files/BP6_Power_sector_maintxt.pdf
4. Statistics South Africa (Stats SA). http://beta2.statssa.gov.za/
5. Statistics South Africa (Stats SA) (2012) General household survey report (P0318). http://www.statssa.gov.za/publications/P0318/P0318August2012.pdf
6. South African Department of Energy (DoE) (2013) Integrated resource plan for electricity (IRP) 2010–2030, update report 2013. http://www.doeirp.co.za/content/IRP2010_updatea.pdf
7. Eberhard A, Kolker J, Leighland J (2014) South Africa's renewable energy IPP procurement program: success factors and lessons, May 2014. http://www.gsb.uct.ac.za/files/PPIAFReport.pdf
8. National Development Plan—Vision for 2030 (2011) http://www.npconline.co.za/medialib/downloads/home/NPC%20National%20Development%20Plan%20Vision%202030%20-lo-res.pdf
9. World Bank (2014) World Bank open data. http://data.worldbank.org/
10. South African Department of Energy (2014) INEP (Integrated national electrification programme). http://www.energy.gov.za/files/policies/p_electricity.html
11. Creamer T (2014) Eskom weighs gas options as diesel costs double to R10.5bn. http://tinyurl.com/nkhu5ah
12. National Climate Change Response White Paper (2013) Department of environmental affairs, navigant research. http://www.navigant.com/~/media/WWW/Site/Insights/Energy/Renewable%20Energy%20Quarterly%20Dialogue%201Q13.ashx
13. KPMG (2011) Carbon disclosure project South Africa's carbon chasm. https://www.cdp.net/CDPResults/south-africa-carbon-chasm.pdf
14. South African Department of Environmental Affairs (DEA) (2010) National climate change response green paper, 2010. http://www.environment.co.za/documents/legislation/south-africa-national-climate_change_response-greenpaper.pdf
15. Briefing to Energy Parliamentary Portfolio Committee. Accessed 1 July 2014. http://www.pmg.org.za/files/140701sanedi.pdf
16. EIA (Energy Information Administration) (2010) Updated capital cost estimates for electricity generation plants
17. IEA (International Energy Agency) (2010) Projected costs of generating electricity. Paris, France

Chapter 4
Energy Generation and Usage in South Africa

Abstract This Chapter focusses on the energy generation and landscape in South Africa. An Energy Sankey Diagram for the Republic of South Africa based on 2012 energy data is also shown.

Keywords Energy generation · South africa · Coal · Energy sankey diagram

Total energy consumption in RSA is ~1100 TWh/year, with usage percentages shown in Fig. 4.1 (2009).

Within this, the mining sector represents 18 % of the country's GDP (8.6 % direct, 10 % indirect and induced), also consumes 2–6 % of all power produced; however, a recent document from *Eskom* stipulates up to 15 % of the utility's annual output is used for mining [2].

The sector is acutely aware of this and the major mining houses are making concentrated efforts to reduce their power consumption. The "big three" platinum miners, *Anglo Platinum*, *Impala* and *Lonmin*, for example, are working to reduce their energy consumption, though as of yet when the three sets of energy consumption data are combined any reduction is masked and cannot be seen.

In terms of production, RSA is one of the most coal dependant countries in the world, using it for 85 % of electricity generation. While it is largely used to generate electricity, other sectors make heavy use of coal as well: 43 % of industrial demand 32 % of commercial and agricultural, and 28 % of residential demand are met directly by coal [3]. When the coal content of the electricity used in these sectors is counted, it provides 80 % of industrial, 80 % of commercial and 57 % of residential demand. In addition to this, a significant amount of coal is channelled to synthetic fuel and petrochemical operations: ~30 % of RSA's vehicle fuel is produced by coal-to-liquids, with an efficiency of ~50 % [4]. Finally, ~28 % of coal production is exported. These trends are very clearly shown in the Energy Sankey Diagram (Fig. 4.2).

© The Author(s) 2016
B.G. Pollet et al., *The Energy Landscape in the Republic of South Africa*,
SpringerBriefs in Energy, DOI 10.1007/978-3-319-25510-1_4

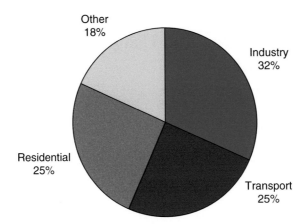

Fig. 4.1 RSA energy usage by sector, 2009. *Source* South African Department of Energy [1]

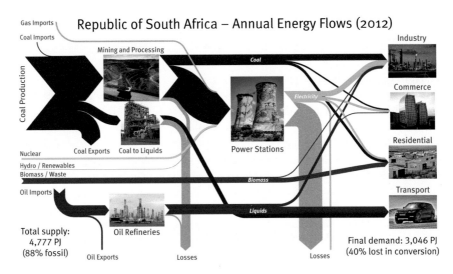

Fig. 4.2 Republic of South Africa *Energy Sankey Diagram. Arrows* show the flow of energy from source through conversion to end use, the width of each arrow is proportional to the annual consumption in 2012. Data from the IEA [3]

According to the historical energy provided in the BP Statistical Review of World Energy, the dominance of coal in RSA has reduced slightly from 75 % of total energy consumption in 1994 to 72 % in 2014. Over the same period, energy consumption has increased from 90 to 122 million tonnes of oil equivalent in 2014 (Fig. 4.3).

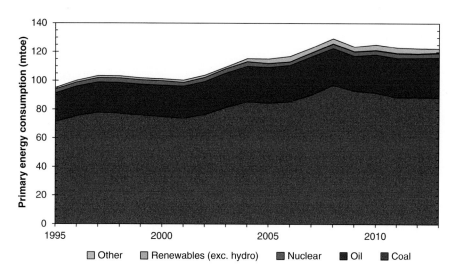

Fig. 4.3 RSA historical energy use, 1994–2014. *Source* BP [5]

References

1. South African Department of Energy (DoE) (2013) Integrated Resource Plan for Electricity (IRP) 2010–2030. Update Report 2013. http://www.doeirp.co.za/content/IRP2010_updatea.pdf
2. http://www.eskom.co.za/sites/idm/Documents/121040ESKD_Mining_Brochure_paths.pdf
3. IEA (International Energy Agency) (2014) World energy statistics and balances (2014 edition)
4. National Petroleum Council (NPC) (2007) Coal to liquids and gas—Topic Paper #18
5. BP (2014). BP Statistical Review of World Energy, June 2014 http://www.bp.com/content/dam/bp/pdf/Energy-economics/statistical-review-2014/BP-statistical-review-of-world-energy-2014-full-report.pdf

Chapter 5
Structure of the Energy Sector in South Africa

Abstract This Chapter highlights renewable energy developments and the various energy initiatives. It briefly discusses the international perspective on capacity shortages as well as the sources of economic damage.

Keywords Renewable energy · REIPPP · South African energy initiatives · Economic damages

5.1 South Africa's Renewable Energy Independent Power Producer Procurement Programme (REIPPPP)

The emergence of renewable energy in RSA has had a long gestation period, but fairly short birth [1]. The country has now become one of the world's largest centres for renewable energy developments, with over 3900 MW of renewable energy projects under construction at a staggering price tag of ~US$14 billion; most of which sourced from foreign and private investors such as Nedbank (involved in 23 projects), Standard Corporate & Investment Banking (17), ABSA (14) Rand Merchant Bank (11) and Investec (4) [2, 3].

Currently the process is set up to procure renewable energy through bid windows for Independent Power Producers (IPPs). Each window provides a maximum number of MegaWatts of power that can be purchased for a set price, differentiated by technology. Once these are sold it is then the responsibility of the IPP to provide the capacity. If this does not yet exist, as in most cases, the IPP must build out the renewable energy infrastructure. Africa's power pool suggests that the continent has immense resources such as gas, coal, hydropower and an abundance of renewable energy sources. Some remarkable successes in certain energy sectors have been noted, such as the RSA's REIPPPP, in which the private sector has heavily invested

B.G. Pollet et al., *The Energy Landscape in the Republic of South Africa*,
SpringerBriefs in Energy, DOI 10.1007/978-3-319-25510-1_5

into renewable energy technologies (see later). In terms of this 20-year projection plan on electricity demand and production, ~42 % of electricity generated should in theory originate from renewable resources. The RSA government has recognised that Energy is an important catalyst for economic growth and increased social equality with a focus in investing in energy infrastructure. In fact, there exists strong opportunity for RSA to provide energy solutions to the rest of Africa.

The REIPPPP is aimed at bringing additional Megawatts into the existing electricity infrastructure through the private sector investment in wind, solar, photovoltaic (PV), concentrating solar power (CSP), biomass, biogas, and small hydro technologies. These renewable technologies form part of the broader energy mix that includes coal, gas, nuclear and imported hydro-energy technologies intended to be used to meet RSA's growing energy demand. To date the RSA DoE has committed to purchase over 2400 MW from IPPs under Windows 1 and 2 of the REIPPPP. Most of the projects under Windows 1 and 2 are currently under construction.

Since the IRP was initiated in March 2010, the RSA DoE has entered into 28 agreements under Bid Window 1 (November 2012) and 19 agreements under Bid Window 2 (May 2013). A further 93 bids were received under bid Window 3 (August 2013). The bids amount to 6023 MW, whilst the available MW for allocation is 1473 MW.

RSA is presently rated as the 12th most attractive investment destination for the implementation of renewable energy technologies [4]. The programme has to date attracted over US$14 billion in foreign direct investment. This bodes well for RSA, as the programme has received international acclaim for fairness, transparency and the certainty of its programme [5]. A progressive increase in local content and job creation numbers has also been witnessed. Window 3 will contribute US$0.4 billion to socio-economic development, aggregating to a cumulative investment of US$0.8 billion. So far the state-owned Industrial Development Corporation (IDC) has approved US$1.24 billion in support for green projects.

Following the three competitive bidding rounds under South Africa's REIPPPP, 64 projects (generating a total of 3922 MW from grid-connected wind, PV and CSP as well as hydro, landfill gas and biomass) at different sites have been selected to proceed, of which 47 are either under construction or already operational as of November 2014. Private sector investment is to date totalling US$14 billion and the projects have attracted investments of over US$8.8 billion.

The RSA government has secured commitments for the supply of 3725 MW of renewable energy by 2016 as a first step to realising the objectives under the IRP2010-2030. It is understood that a total of 1105 MW has been allocated for the 4th bid Window, divided between onshore wind (590 MW), PV (400 MW), biomass (40 MW), landfill gas (15 MW) and small hydropower (60 MW).

The bid windows will continue annually until the required amount of renewable energy is secured. The results from the first 3 bid windows are summarised in Table 5.1, with detailed data given in Appendix A. However, some recent concerns have been expressed and issues have been raised with some of the projects (26) being connected to the grid run by *Eskom*.

Table 5.1 Key results from the South African REIPPPP Bid Windows 1, 2 and 3. *Source* World Bank [6] and Eberhard [2]

	Wind	PV	CSP	Hydro	Biomass	Biogas	Landfill	Total
Capacity offered (MW)								
Bid Window 1	1850	1450	200	75	12.5	12.5	25	3625
Bid Window 2	650	450	50	75	12.5	12.5	25	1275
Bid Window 3	654	401	200	121	60	12	25	1473
Capacity awarded (MW)								
Bid Window 1	634	632	150	0	0	0	0	1416
Bid Window 2	563	417	50	14	0	0	0	1044
Bid Window 3	787	435	200	0	16	0	18	1456
Projects awarded								
Bid Window 1	8	18	2	0	0	0	0	28
Bid Window 2	7	9	1	2	0	0	0	19
Bid Window 3	7	6	2	0	1	0	1	17
Totals from all windows								
Capacity awarded (MW)	1984	1484	400	14	16	0	1	3915
Average success rate	63 %	64 %	89 %	5 %	19 %	0 %	24 %	61 %
Project awarded (MW)	32	23	5	2	1	0	1	64
Investment (US$ millions)	4683	5085	3806	79	108	0	29	14,011

Table 5.1 clearly shows that the learnings from each bid window have been steep. One of the key learnings was the move to annual bid windows, not twice yearly, as was the case for the first two. Also awarding a smaller number of projects appears to be preferable. In total, the value of the first three bid windows was ~US$14 billion. The 3916 MW awarded so far is only a small part of the 20,000 MW by 2030, but is on track to reach that target if the current momentum can be maintained. A Round 4 tender has started in August 2014 as after Round 3, 2808 MW still needs to be allocated which comprise 1041 MW of PV, 1336 MW of wind, 200 MW of solar CSP, 121 MW of small hydro and 110 MW of biomass, biogas and landfill gas.

5.2 Other South African Energy Initiatives

The **Green Economy Accord** identifies the green economy as one of the 12 job drivers that could help creating 5 million additional jobs by 2020 [7]. Indeed, the Accord identifies 12 areas of commitment for partnership between government and social partners under the green economy. The Industrial Development

Corporation [8] recently projected that RSA could create direct green jobs from natural resource management, green energy generation, energy and resource efficiency, and emissions and pollution mitigation. The report projects the creation of 980,000 jobs in the short term, between 2011 and 2012, 255,000 jobs in the medium term, between 2013 and 2017, and 462,000 jobs between 2018 and 2025 in the long term.

Currently, Africa's manufacturing sector accounts for only 15 % of the overall yearly export earnings whilst in RSA, manufacturing contributes between 10 and 15 % a year to the economy. This contrasts greatly with China, where the manufacturing sector contributes up to 50 % to the country's annual GDP. To date, the **Green Fund** (www.greenfund.org.za) has received a total of 590 funding applications for potential sustainable-development projects amounting to US$0.96 billion. This fund is purely dedicated to create and stimulate Green Economy.

The **Integrated National Electrification Programme (INEP)** initiated by the RSA DoE has allocated over US$0.26 billion (FY2013/14) to projects. The INEP has the purpose to frontload loan facility to enable low-capacity municipalities in disadvantaged regions to fast-track the supply of electricity to their communities. The programme formed part of government's plan to accelerate the eradication of electricity backlogs to meet the 2015 Millennium Development Goals (MDGs).

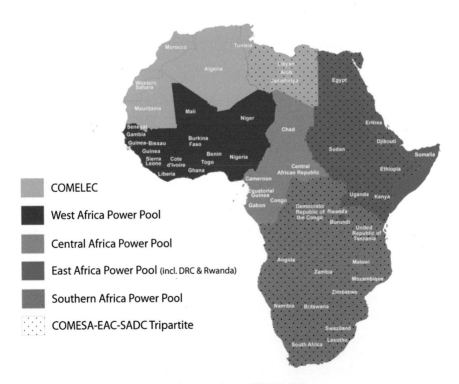

Fig. 5.1 Southern Africa power pool region. *Source* IRENA [10]

Table 5.2 Southern African power pool production by utility and fuel: 2012. *Source* Navigant Research, 2013 [11]. CCGT stands for combined cycle gas turbine

Country	Utility	Hydro/ MW	Coal/MW	Nuclear/ MW	CCGT/ MW	Distillate/ MW	Total/ MW
South Africa	*Eskom*	2000	37,831	1930	–	2409	44,170
Mozambique	*EDM*	2573	–	–	–	51	2624
DR Congo	*SNEL*	2442		–	–	–	2442
Zimbabwe	*ZESA*	750	1295	–	–	–	2045
Zambia	*ZESCO*	1802				10	1812
Angola	*ENE*	760	267	–	160	–	1187
Tanzania	*Tanesco*	561	–	–	485	78	1124
Namibia	*NamPower*	240	132	–	–	21	393
Malawi	*Escom*	286	–	–	1	–	287
Botswana	*BPC*	–	132	–	–	70	202
Lesotho	*LEC*	72	–	–	–	–	72
Swaziland	*SEC*	63	9	–	–	–	72

Development Bank of Southern Africa (DBSA) has recently announced that it would inject around US$35 million into the programme to fast-track the implementation of the electrification of >20,000 households in rural regions (e.g. Limpopo, the Eastern Cape and KwaZulu-Natal) [9].

RSA is also part of the **Southern African Power Pool (SAPP)**, a consortium of 12 utilities operating across 12 neighbouring countries in Africa (Fig. 5.1), namely: Angola, Botswana, DRC, Lesotho, Malawi, Mozambique, Namibia, Swaziland, Tanzania, Zambia, and Zimbabwe [10].

In 2012 those 12 countries constitute a current capacity of only 54 GW to provide power to a combined population of 189 million people. In contrast, the UK has a capacity of around 90 GW to supply just 62 million people. The aim of SAPP is to create a common market for electricity with all utilities acting without taking advantage of one another, thereby allowing a faster growth of a common power market across the region.

Table 5.2 shows the current utilities signed up in the SAPP and the 2012 data for their power production, broken down by fuel type. This clearly shows the dominance of *Eskom* over the entire region, its disproportionate usage of coal and distillate, and its status as sole user of nuclear power.

5.3 International Perspective on Capacity Shortages

Society's absolute dependence on electricity means that disruptions to supply will have severe social and economic consequences. The growing rate of electrification and consumption per capita in RSA means that the costs of electricity outages will only continue to grow.

Power outages damage a country's economy in four ways [12, 13]:

- Indirect costs from lost time and productivity;
- Direct costs from lost assets, such as damaged production lines or spoiled food;
- Loss of investor confidence leading to an out-flow of industry;
- Long-term increase in electricity prices, increasing the burden on industry and households.

There is a rich literature on the cost of electricity shortfalls since the oil crisis of the 1970s [12], with numerous estimates for individual blackouts and ongoing capacity shortages from around the world.

5.3.1 Sources of Economic Damage

5.3.1.1 Loss of Productivity

Loss of output is by far the largest source of economic damage [14]. Industry, commerce and government services are increasingly reliant on uninterrupted electricity supply. The obvious examples are hospitals and high-tech manufacturing, but with the spread of just-in-time supply chains, the disruption of one supplier of even a basic component or service can propagate upwards and delay the entire manufacturing process. In the wider picture, power outages can disable public transportation systems, the ability to perform useful tasks at work or at home, incurring costs from lost time and productivity.

Planned outages are widely regarded as being less damaging than unexpected outages, as they give the opportunity to reschedule activities which are critically dependent on electricity. However, planned outages are still a major inconvenience, as processes and production lines can be costly and slow to restart, whereas those that are designed to run 24/7 have to be closed down, reducing productivity.

5.3.1.2 Loss of Assets

Unexpected outages can, on the other hand, destroy goods and assets. For example, during the 2003 blackout in Northeast US and Canada, Daimler Chrysler lost production at 14 plants. Around 10,000 cars were moving through paint shops at the time of the outage and had to be scrapped due to damage. At one Ford plant, the outage caused a molten car shell to solidify inside one of their furnaces, putting it out of action for weeks [15]. The catering industry is particularly susceptible due to the need for cooling—an outage of several hours without backup can cause food stocks to perish, causing both economic loss, and if they persist for days, the much larger problem of physical shortages of sustenance.

5.3.1.3 Loss of Confidence

Investors naturally seek reliable environments in which to locate their facilities. Many companies are unwilling to even source from suppliers located in countries with unstable power supplies, due to the knock-on impacts they may have on the supply chain [12]. If a country is unable to maintain a reliable electricity supply, it follows that the mobile parts of its industrial sector will begin to relocate, and that global companies will be deterred from building new facilities.

The perception of insecurity is broader than just the loss of electricity supply. It can be taken as a signal that the country has a deeper structural problem: a government that is unable to manage its electricity infrastructure, may also struggle to control other vital services. Francois Stofberg is quoted in Bloomberg: "The impact of Eskom's shortages on production has been underestimated. The question is how long it will be before investors look at our growth rate and decide to take their investments, and especially their capital investments, elsewhere." [16].

In a survey of industrial leaders, security of energy supply was noted to be a top priority when deciding where to site new facilities [12]. Companies routinely incorporate the cost of back-up power when looking at less developed countries, decreasing their competitiveness against countries with reliable infrastructure (such as those in South East Asia).

5.3.1.4 Increase in Prices

In the short term, industry and households respond to unreliable electricity supply by investing in their own backup systems. During Nigeria's electricity crisis of the 1990s, nearly three-quarters of manufacturers owned some form of backup generation, adding 20–30 % to the upfront cost of establishing new facilities [14]. Compounding this was the fact that electricity from these backup systems cost 2–3 times more than from the public supply.

As the difference between installed capacity and peak demand (the reserve margin) decreases, long term electricity prices rise exponentially, and can "take astronomical values" as it falls towards zero [17]. This is currently being experienced in South Africa, as expensive diesel generators are being called on for a large portion of the year, dramatically increasing the operating costs of the electricity system.

5.3.2 Estimates for the Cost of Outages

Classic studies from the 1980s and 1990s estimated the cost of unreliable electricity supply at 1–2 % of national GDP in India, Pakistan and Columbia [18, 19]. More recent estimates suggest that GDP growth rates in Sub-Saharan Africa are held back by 2 percentage points because of the weak power infrastructure [20].

Two recent examples from the US provide point estimates for the cost of power system failures. The capacity crisis suffered by California in 2000/01 resulted in 1.5 million people enduring rolling blackouts for nearly a year. This was estimated to add ~$40 billion to energy costs during 2001–03 [21] and reduce the state's GDP by around 1 % [22]. Retail prices increased by 35 %, after the crisis. A system failure in the Northeast US and Canada left 50 million people without electricity during the summer of 2003. The resulting four-day blackout was estimated to have cost around $6 bn [12].

Finally, the Great East Japan earthquake and tsunami of 2011 removed 27 GW of generation from service (30 % of the country's total), notably including the 9 GW Fukushima Daichi nuclear reactor which partially melted down. Most plants were gradually brought back online over the period of a few months, but 45 million people endured rolling blackouts for up to 6 weeks [23]. The economic cost of the power outage are conflated by the wider costs of the natural disaster (which killed around 16,000), but electricity shortages are believed to have caused the largest hindrance to Japan's economic recovery [24].

References

1. Ram B (2006) Renewable energy development in Africa—challenges, opportunities, way forward. South Africa Regional Office, African Development Bank
2. Eberhard A, Kolker J, Leighland J (2014) South Africa's renewable energy IPP procurement program: success factors and lessons, May 2014. http://www.gsb.uct.ac.za/files/PPIAFReport.pdf
3. UNEP (2012) Financing renewable energy in developing countries, drivers and barriers for private finance in sub-Saharan Africa, February 2012. http://www.unepfi.org/fileadmin/documents/Financing_Renewable_Energy_in_subSaharan_Africa.pdf
4. South African Department of Energy (DoE) (2013) More agreements signed for energy generation. http://www.sanews.gov.za/south-africa/more-agreements-signed-energy-generation
5. Creamer T (2014) Global renewables investors warn SA prices becoming 'dangerously low'. Engineering news. http://tinyurl.com/qayey42
6. World Bank (2014) World Bank open data. http://data.worldbank.org/
7. Borel-Saladin JM, Turok IN (2013) The impact of the green economy on jobs in South Africa. S Afr J Sci 109(9/10), Art. #a0033, 4 pp. http://dx.doi.org/10.1590/sajs.2013/a0033
8. Maia J, Giordano T, Kelder N, Bardien G, Bodibe M, Du Plooy P, Jafta X, Jarvis D, Kruger-Cloete E, Kuhn G, Lepelle R, Makaulule L, Mosoma K, Neoh S, Netshitomboni N, Ngozo T, Swanepoel J (2011) Green jobs: an estimate of the direct employment potential of a greening South African economy. Industrial Development Corporation, Development Bank of Southern Africa, trade and industrial policy strategies. http://www.idc.co.za/projects/Greenjobs.pdf
9. DBSA (Development Bank of Southern Africa) (2014) Integrated annual report. http://issuu.com/developmentbankofsouthernafric/docs/dbsa_integrated_annual_report_2013
10. IRENA (International Renewable Energy Agency) (2013) Southern African power pool: planning and prospects for renewable energy. http://www.irena.org/DocumentDownloads/Publications/SAPP.pdf
11. National Climate Change Response White Paper (2013) Department of environmental affairs, navigant research. http://www.navigant.com/~/media/WWW/Site/Insights/Energy/Renewable%20Energy%20Quarterly%20Dialogue%201Q13.ashx

12. Royal Academy of Engineering (2014) Counting the cost: the economic and social costs of electricity shortfalls in the UK
13. Praktiknjo AJ, Hähnel A, Erdmann G (2011) Assessing energy supply security: outage costs in private households. Energy Policy 39:7825–8833
14. Adenikinju AF (2003) Electric infrastructure failures in Nigeria: a survey-based analysis of the costs and adjustment responses. Energy Policy 31:1519–1530
15. Electricity Consumers Resource Council (2004) The economic impacts of the August 2003 blackout. Electricity Consumers, Washington DC
16. Bloomberg Business (2015) South African power outages hurt economy as unemployment climbs. http://tinyurl.com/q6o8oy5
17. Amobi MC (2007) Deregulating the electricity industry in Nigeria: lessons from the British reform. Socio-Econ Plann Sci 41:291–304
18. USAID (1988) Power shortages in developing countries: magnitude, impacts, solutions, and the role of the private sector
19. Kessides C (1992) The contributions of infrastructure to economic development: a review of experience and policy implications. World Bank Discussion Paper 213
20. Andersen TB, Dalgaard CJ (2013) Power outages and economic growth in Africa. Energy Econ 38:19–23
21. Weare C (2003) The California electricity crisis: causes and options. Public Policy Institute of California, San Francisco
22. Cambridge Energy Research Associates (2001) Short circuit: will the California energy crisis derail the state's economy?. UCLA, Los Angeles
23. Ministry of Economy, Trade and Industry (2011) Economic impact of the Great East Japan earthquake and current status of recovery, Tokyo
24. Swedish Agency For Growth Policy Analysis (2011) Restarting Japan: a first assessment June 2011 on the road to recovery after the Great East Japan earthquake. Working paper 2011/15

Chapter 6
Past, Current and Future Energy Production

Abstract This Chapter focusses on various forms of electricity generation: coal, natural gas, nuclear and renewables. It also discusses the future energy production including Hydrogen Energy.

Keywords Renewable energy · Electricity generation · Coal & gas · Hydrogen energy

6.1 Coal: Past, Present and Future

Coal was the primary ingredient which sparked the industrial revolution in Western societies 200 years ago [1]. Global coal consumption is expected to rise by 25 % by the end of the decade to 450 billion tonnes of oil equivalent, overtaking oil at 440 billion tonnes [2]. Half of China's power generation capacity to be built between 2012 and 2020 will be coal-fired [2].

RSA's abundance of coal, present lack of liquid fuel reserves, and the historic isolation of the country have created a situation where coal is the primary source of energy. By looking to the history of other countries, the future of RSA's coal production can be better understood. Future projections are prone to falling into the trap of assuming that current conditions will prevail. This can lead to over-optimism, predicting that current growth can carry on indefinitely, or the expectation of a cataclysmic crash: projecting that current reserves will "run out" after a fixed number of years based on current production and stocks. In reality, resource extraction changes very gradually from year to year, except during major external disturbances such as wars or industrial strikes.

Figure 6.1 shows the history of coal consumption in two prominent countries (RSA and UK), showing how the UK is at the end of its coal era, while RSA

© The Author(s) 2016
B.G. Pollet et al., *The Energy Landscape in the Republic of South Africa*,
SpringerBriefs in Energy, DOI 10.1007/978-3-319-25510-1_6

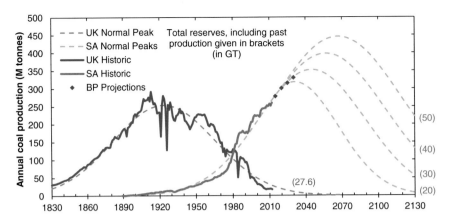

Fig. 6.1 Three hundred years of historic and projected coal consumption in the UK and South Africa. Data from [3, 4]

production is still rapidly expanding. After a 100 years of steady growth, the UK's coal production peaked in the early twentieth century, and has since been in gradual decline. Annual production now stands at 17 MT per year, one twentieth of its peak. The UK dug up 80 % of its economically viable coal between 1870 and 1970, and is now left with just 1 % of its reserves left underground: 0.2 of 27.6 GT [3, 4].

In contrast, RSA's coal production has grown steadily by 2 % per year since the early 1990s, and around three quarters of its viable coal reserves remain underground: 30.1 of 38.6 GT total (including historic production). A naive assumption is that with 260 MT annual production in 2012, RSA coal reserves will last for 115 years. This neglects both the increasing rate of production, and improvements in technology that allow more coal to be found and extracted at a given price [5].

It is possible to speculate about the future trajectory of production in RSA, using a variation on the Hubbert's Peak theory used to assess peak oil. Several future curves are presented in Fig. 6.1, based on different estimates for the country's total reserves, which have ranged from 15 to 55 GT in the past decade, and currently stand at ~30 GT [4]. Peak coal in RSA can be expected to occur between 2031 and 2067 depending upon the recoverable reserves.

The future projections are based on Hubbert's Peak theory, which fits a normal distribution to historic consumption, with the integrated area under the curve equal to the total amount of resource that can be extracted. Hubbert's theory uses the Ultimately Recoverable Resource (URR), encompassing all geological deposits regardless of whether they are technically or economically feasible to exploit, on the underlying assumption that eventually, technology improvement and price increases will mean that everything will be exploited. Figure 6.2 instead follows Rutledge [3] in using coal reserves, which are just those resources which can be profitably extracted with current technology. This stems from the enormous over-estimates of British coal reserves made during the *Victorian era*, leading to

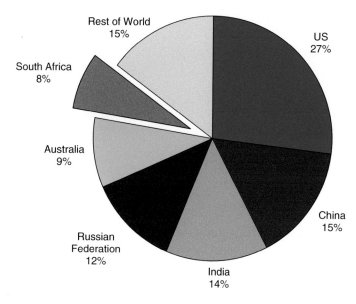

Fig. 6.2   Proven Reserves of Anthracite and Bituminous Coal, by Top 6 Countries, 2013. *Source* BP [4]

predictions of eleven centuries of bountiful supply (which lasted for only two). While the UK still has significant amounts of coal, it is trapped in seams under a metre thick, more than 1,000 metres below ground, meaning it is much cheaper to import coal from abroad.

RSA's proven reserves amount to ~8 % of total global coal reserves (see Fig. 6.2). However, if the international ambition to limit global warming to 2 °C is acted on, then an estimated 85 % of the coal reserves in Africa would have to remain unburned—amounting to 28 GT across the continent, and 25.6 GT in RSA alone [6].

6.2   Natural Gas

To date natural gas in Africa has been a very minor fuel [7]. This is likely to change over the next decade as gas has the potential to account for more than 40 % of the electricity generated in sub-Saharan Africa (SSA) from 2020 onwards, a new report by McKinsey & Company shows, adding that, by 2040, gas-fired capacity could be responsible for more than 700 terawatt-hours in the region [8]. In RSA this lack of usage and interest is clearly shown in a map of gas infrastructure in Fig. 6.3.

Although an increasing focus is being given to potential reserves of shale gas in the Karoo Basin, and the RSA government has now identified shale gas as a

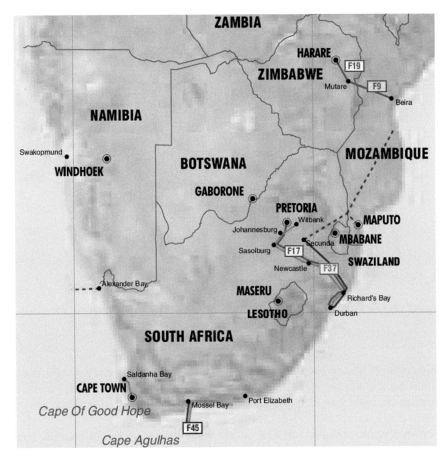

Fig. 6.3 Map of Southern Africa constructed and planned natural gas pipelines *Source* http://www.theodora.com

"Potential Game changer" [9]. RSA's lower Karoo is estimated to contain 1,559 trillion cubic feet (tcf) of potential shale gas reserves, of which 390tcf is classed as (currently) technically recoverable [10]. Putting this into context, 390tcf has the equivalent energy content of around 13.2 GT of coal, adding around a third to RSA's total reserves of fossil energy. This puts RSA's potential shale reserves in the top ten in the world, as shown in Table 6.1.

At the time of writing, exploration licenses have been granted to *Shell International, Falcon Oil and Gas* in partnership with *Chevron*, and *Bundu Gas*. Further exploration licenses are expected to be awarded throughout 2015.

Any potential exploitation of the reserves is controlled under the 2014 amended "Mineral and Petroleum Resources Development Act", which falls under the RSA Department of Minerals (DMR). The updated act gives the government a 20 % stake in all developments, and the right to purchase up to 80 % of the remaining,

Table 6.1 Current estimates of recoverable shale reserves of top 10 countries

Country	January 1, 2013 estimated proven natural gas reserves (tcf—trillion cubic feet)	2013 EIA/ARI unproved wet shale gas technically recoverable resources (TRR) (tcf—trillion cubic feet)
China	124	1,115
Argentina	12	802
Algeria	159	707
Canada	68	573
United States	318	567
Mexico	17	545
Australia	43	437
South Africa	–	**390**
Russia	1,688	287
Brazil	14	245

Source Energy Information Administration [10]

at a pre-agreed upon price. This in effect means that the reserve would be government controlled.

In summary natural gas, specifically shale gas, is of increasing interest in RSA, but too date remains unexplored and unexploited. The time it will take before the first native RSA gas is pumped though remains questionable. The National Gas Infrastructure Development (NGID) is due for release at some point during 2015.

6.3 Electricity Generation

Figures 6.4 and 6.5 show the make-up and geographical distribution of RSA's power station infrastructure, based on detailed data given in Appendix A. These clearly show that coal will remain key to RSA's energy landscape for a long time to come. Due to the dependency on coal, carbon emissions from the power sector in RSA are a key focus of attention.

Investment in new generating capacity has failed to keep pace with increasing demand. This has led to dramatic cost escalation as the available capacity has been stretched to its limits and forced to run beyond its intended utilisation. In 2013 *Eskom*'s two open-cycle gas turbines (OCGT) produced 3,621 GWh—translating to a capacity factor of over 19 % [12]. The usage of these plants has quadrupled in the space of 2 years, and now stands well above the intended utilisation of 2–4 %. During 2013, these two OCGT plants consumed ~US\$900 million in diesel fuel, translating to a levelled cost of energy of ~US\$250/MWh—approximately five times higher than if this demand could have been met by additional baseload coal capacity [12].

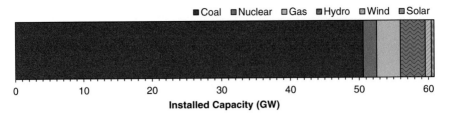

Fig. 6.4 Installed power plants in RSA by type. *Source* see data in Appendix A. *Source* http://www.theodora.com

Fig. 6.5 Map of power plants in RSA. Marker size is proportional to capacity (500 MW shown in the legend). Data from Davis [7] and Pierrot [11]

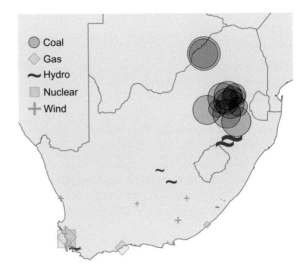

The 'controlled' blackouts worsened during late 2014 and early 2015 which has severely impacted RSA's economy [since the beginning of 2015 up to now (17/04/15), RSA has experienced 26 days of load shedding so far]. In an attempt to manage the energy demand and supply, *Eskom* introduced and implemented load-shedding stages (3 in total). Stage I allows for up to 1 GW, stage II for up to 2 GW and stage III for up to 4 GW of the national load to be shed. According to RSA government's energy advisors, this stages of load shedding has cost the economy (based on 10 h of blackouts per day for 20 days per month) US$1.6 million per month for Stage I, US$3.2 million per month for Stage II and US$6.6 million per month for Stage III. Very recently (April 2015), *Eskom* announced that it will be forced into rolling blackouts throughout 2015 (mainly due to ongoing maintenance issues, the shortages of diesel and the collapse of one of its coal storage silos. According to the JSE-listed financial services company Efficient Group these power cuts has so far cost RSA's economy > USD$25 billion (since 2007) equating to more that 1 million job opportunities.

6.4 Future Energy Production

The key document, in terms of deciding future energy policy, is the IRP2010–2030. Using the current 2012 update, we can see that by the end of 2012 there was a committed new build programme of 14.8 GW of energy generation from 2010 to 2020. The types of power generation are shown in Fig. 6.6.

All of the coal new build identified in the IRP is through *Eskom*. Taking into account its current financial difficulties it is likely that some of the coal new build will slip, or could be reassigned, potentially, to natural gas.

An update of the IRP is due soon (with a publication scheduled for the end of 2014) and should take into account the natural gas plan. Firstly, the implications of this are that this data, out to 2020, will change probably with a large shift from coal to gas. Secondly *post-2020* RSA is working on a scenario based approached, this includes a move to large scale deployment of nuclear, renewables and the usage of power from the Grand Inga Dam Project.

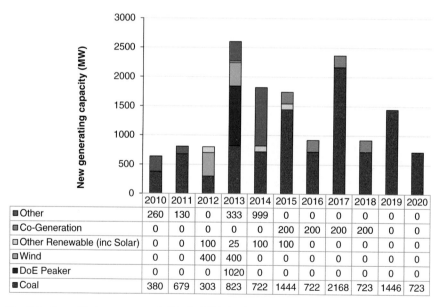

	2010	2011	2012	2013	2014	2015	2016	2017	2018	2019	2020
■ Other	260	130	0	333	999	0	0	0	0	0	0
▣ Co-Generation	0	0	0	0	0	200	200	200	200	0	0
▫ Other Renewable (inc Solar)	0	0	100	25	100	100	0	0	0	0	0
▣ Wind	0	0	400	400	0	0	0	0	0	0	0
■ DoE Peaker	0	0	0	1020	0	0	0	0	0	0	0
■ Coal	380	679	303	823	722	1444	722	2168	723	1446	723

Fig. 6.6 Committed energy new builds in RSA for 2010–2020. *Source* South African Department of Energy [13, 14]

6.4.1 Carbon Capture and Storage (CCS)

Due to the countries reliance on coal it is not surprising that the RSA government, in accordance with its commitments to reduce greenhouse gases, has decided to move along the path of Carbon Capture and Storage. The country's CCS roadmap, which is optimistic, has ambitious key dates [15]. RSA is using the significant Norwegian competences in the field of CCS to further its own projects. This forms part of the bilateral agreement between the countries, which also includes renewable energies. Norway's National Petroleum Directorate has pledged US$2.5 million to support RSA's pilot CCS project, in addition to the US$25 million allocated by the World Bank.

6.4.2 Nuclear

Around 5–6.5 % of RSA's electricity is provided via *Eskom*'s 1,800 MW Koeberg Nuclear Power Station's two reactors (French-built) in the Western Cape Province. The Nuclear Energy Policy (NEP) set by the state-owned Nuclear Energy Corporation of South Africa (NECSA) aims to (i) increase the role of nuclear energy as part of the process of diversifying RSA's primary energy sources to ensure energy security, (ii) reducing the country's over-reliance on coal and (iii) in the long-term vision for RSA to become globally competitive in the use of innovative technology for the design, manufacture and deployment of state-of-the-art nuclear energy systems and power reactors, and nuclear fuel-cycle systems. As part of the Energy plan, it was recently announced that RSA will build six new nuclear power plants by 2030, providing 9,600 MW (9.6 GW) of power at a cost estimated between US$36 billion and US$90 billion. This causes great concern as the current national debt is estimated at ~US$140 billion (~40 % of the GDP) and the RSA's credit rating could be threatened if it commits itself to nuclear installations it cannot afford. These figures are quite high and for comparison purposes, even the most expensive plant in the world such as the EDF's proposal for Hinkley Point C in the UK is estimated at ~US$5,000/kW—which would amount to ~US$48 billion for these planned plants. This cost could indeed be substantially lower if the plants were built by a government owned utility (such as *Eskom*) which could borrow for less than a 10 % interest rate, greatly reducing the cost of financing the investment. Recently the RSA government signed several Inter-Governmental Agreements with USA, South Korea, Russia, France and China with the intention of procuring and selecting strategic partner(s) to carry on the nuclear build programme (see IRP 2010–2030).

6.4.3 Renewables

In RSA, there is an increase in the installation of solar water heaters and solar PV panels in commercial buildings and private dwellings. However, it is nowhere near the levels of installation seen in the countries in Europe, the US or Japan. This may seem surprising, as these countries have far lower levels of solar irradiance; however they have much higher GDP per capita, meaning that these technologies are more widely affordable despite producing less energy. Southern Africa has one of the highest numbers of sunny days in the world (more than 2,500 h of sunshine/ year) and significantly higher levels of radiation (average solar radiation levels ranging between 4.5 and 6.5kWh/m^2/day) [16].

Solar technologies, particularly concentrating solar power (CSP) are rapidly approaching grid parity in RSA due both to the abundance of resource and the recent rises in electricity prices [17]. CSP is of particular interest as it can be combined with thermal storage to provide baseload dispatchable power [18]. Gauche argues that CSP is the only sustainable and dispatchable technology that could supply South Africa's electricity demand, and that "*a balanced mix of PV, wind and CSP can provide the energy supply needed in South Africa*" [19].

The wind resource in South Africa is also particularly strong, especially along the southern half of the country. Figure 6.7 depicts the long-run average wind speeds across the country, overlaid with electricity network infrastructure. Much of the country benefits from average wind speeds in excess of 5 m/s (often considered the minimum requirement for commercial viability), and significant areas are on a par with the best sites in Europe and the US with speeds over 7.5 m/s.

The pattern of output from wind farms is of particular interest due to its unpredictable and undispatchable nature. Historic data from RSA is not available, in part because wind power has only recently taken off in the country.

In lieu of actual data, the Virtual Wind Farm model was used to simulate the pattern of output that could be expected from a distributed set of wind farms, shown as circles in Fig. 6.7. The model relies on historic weather data from NASA, converting hourly wind speeds at 2, 10 and 50 m above ground at regularly spaced observation points into the estimate power output at the specific location and height of specific wind farms. It is described more fully in [21] and validated for the UK in [22]. At each site, a farm of thirty Vestas V90 3 MW turbines was modelled, to give an indication of the combined output that RSA could expect once several of the REIPPPP projects have been developed.

Figure 6.8 plots the simulated output from these farms over the course of 2 years. While the output from each individual farm is quite variable, their geographical dispersion means that the aggregate output of the whole set of farms is more consistent. The output shows a consistent seasonal profile, being lowest in February and March, and highest during June to November.

The 5th and 95th percentile of the fleet-aggregated capacity factors was simulated to be 16 and 30 % respectively, with an average over the period of 2008–10 of 22.1 %. It should be noted that the locations chosen were not formally optimised, but chosen to approximately balance the needs of high annual average wind speed and proximity to existing transmission infrastructure. In practice, South Africa's fleet of wind farms may operate with higher average capacity factors due to developers choosing more optimal locations.

Figure 6.9 summarises the hourly capacity factors that were simulated for the entire fleet. These follow the shape of a Weibull distribution, with a mean of

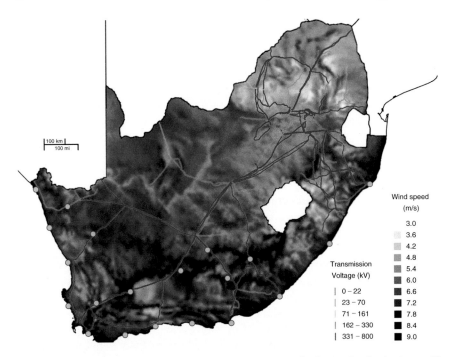

Fig. 6.7 Average wind speeds across South Africa (shown as *background colour*), along with the location of electricity transmission infrastructure (shown as *lines*), and the hypothetical wind farm locations that were used for wind output simulation (shown as *circles*). Data from 3TIER and AICD via IRENA [20]

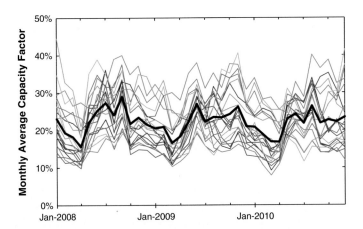

Fig. 6.8 Simulated monthly output from the 20 wind farms distributed across South Africa (shown as *thin lines*), and the aggregated output of those farms (*thick black line*)

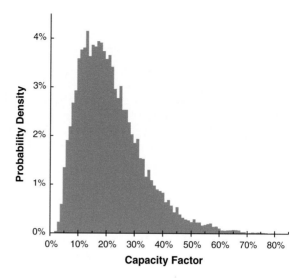

Fig. 6.9 Distribution of hourly capacity factors simulated for the aggregate fleet of 20 wind farms

22.1 % and peak of 82.4 %. The 5th percentile is 7.4 % and the 1st percentile is 5.1 %—highlighting the fact that wind energy is not a firm and reliable source of power. It should be remembered that whilst wind power will certainly reduce the spend on fuel for conventional generation, it will do little to ease the capacity shortage in RSA, as it cannot be relied upon to generate at times of peak demand.

6.5 Mineral Resources and Hydrogen Energy

RSA is well-endowed with Platinum Group Metals (PGM, such as Platinum, Palladium, etc.) and other mineral resources such as Titanium, Vanadium, Chromium, etc. (Figure 6.10). Minerals and PGM beneficiation is currently a top priority for the RSA government as it is hoped to unlock foreign investments and to attract downstream value-adding manufacturers in turns creating jobs. The main focus is to add-value to titanium, iron, steel and platinum. For the latter, the aim is to expand the Auto catalyst sector and to create a Hydrogen and Fuel Cell industry in strategic geographical areas such as the Special Economic Zones (SEZs).

The financier and mining guru Robert Friedland (founder of Africa-focused project developer *Ivanhoe Mines*, formerly known as *Ivanplats*) recently stated that the rise of the hydrogen fuel cell-powered vehicle is about to change the global Platinum landscape forever.

According to the 4th Energy Wave, Fuel Cell and Hydrogen Annual Review [23], in 2014 the global fuel cell industry posted demand for platinum of 25 thousand ounces. This has risen from under 10 thousand ounces in 2013. This jump in usage was due to the increase in PEM stacks for a range of applications. 4th Energy Wave forecasts that in 2015 demand will increase to 34 thousand ounces.

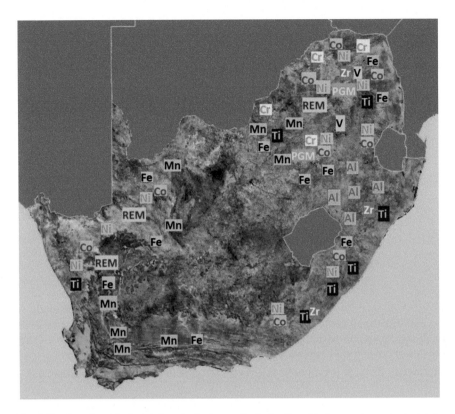

Fig. 6.10 Overview of mineral deposits in the use for fuel cell and hydrogen storage materials in RSA [24]. *PGM* Platinum Group Metals. *REM* Rare Earth Metals

   Looking forward, taking into account thrifting on the one hand and increased demand for low temperature fuel cells (including PEM, DMFC, PAFC and AFC) on the other, the report forecasts that by 2025 platinum demand from the entire fuel cell sector will reach 252 thousand ounces (Fig. 6.11).

   The sharp dip in demand in this chart is created by the release of the next generation automotive PEM stack which is forecast to have significantly lower platinum loadings. This forecast assumes uptake of fuel cell vehicles to be in the tens of thousands by 2025, not in the millions. Luckily South Africa is currently being geared up to a Hydrogen Economy thanks to its Hydrogen and Fuel Cell programme.

   In May 2007, Hydrogen South Africa or HySA was initiated by the RSA Department of Science and Technology (DST) and approved by the Cabinet. HySA is a long-term (15-year) programme within their Research, Development, and Innovation (RDI) strategy, officially launched in September 2008. This National Flagship Programme is aimed at developing South African intellectual property, knowledge, human resources, products, components and processes

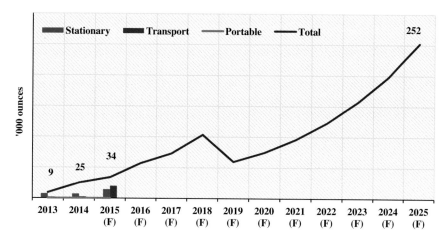

Fig. 6.11 Global platinum demand from the fuel cell sector

to support the South African participation in the nascent, but rapidly developing international platforms in Hydrogen and Fuel Cell Technologies. HySA comprises of three R&D Centres of Competence: HySA Catalysis, HySA Infrastructure and HySA Systems [24].

The programme strives towards a knowledge-driven economy meaning that innovation will form the basis of South Africa's economy; this includes an aggressive capacity-development programme's approach. HySA also focusses on (i) the "Use and Displacement of Strategic Minerals", (ii) ways of harnessing South Africa's mineral endowments to promote both the hydrogen economy and renewable energy use, and (iii) seeking the most cost-effective and sustainable ways of incorporating PGM-based components in hydrogen fuel cell and other technologies, in turns resulting in commercialisation ventures and a viable industry around mineral beneficiation.

Indeed, HySA has been implemented in the context of the DST's various innovation strategies, the Department of Mineral Resources' minerals beneficiation strategy, the Department of Energy's Integrated Resource Plan (IRP, see earlier) and the Department of Trade and Industry's (DTI) industrial development strategies. The principal strategy of HySA is to execute research and development work, with the main aim of achieving an ambitious 25 % share of the global Hydrogen and Fuel Cell market using novel Platinum Group Metal (PGM) catalysts, components and systems since South Africa has more than 75 % of the world's known PGM reserves. In order to achieve this, the structure is aimed at the parallel development of knowledge and technology across all areas of the Hydrogen and Fuel Cell value chain, allowing for the establishment of a strong R&D Hydrogen and Fuel Cell Technology exporting added value PGM materials, components and complete products. Each Centre has a unique responsibility, but all three are complementary within the common vision of fostering proactive innovation and

developing the human resources required to undertake competitive R&D activities
in the field of Hydrogen and Fuel Cell Technologies. The first 5 years of funding
focused on developing infrastructures at each Centre with a major emphasis upon
Human Capacity Development (HCD). Relevant (inter)national expertise was
recruited by each Centre to access technical support and well-established implementation networks, and to ensure the programme and its deliverables remain
market related and world-class. Furthermore, to achieve the HySA strategy objectives, the three HySA Centres of Competence form a national network of research
'Hubs' and 'Spokes' through collaboration with institutions and partners from the
R&D sector, higher education, as well as industry.

References

1. Shafiee S, Topal E (2009) When will fossil fuel reserves be diminished? Energy Policy 37:181–189
2. Mackenzie W (2011) Coal supply service: South Africa: production August 2011 Report. http://woodmacresearch.com. Accessed Sept 2014
3. Rutledge D (2011) Estimating long-term world coal production with logit and probit transforms. Int J Coal Geol 85(1):23–33
4. BP (2014) BP Statistical Review of World Energy. http://www.bp.com/content/dam/bp/pdf/Energy-economics/statistical-review-2014/BP-statistical-review-of-world-energy-2014-full-report.pdf
5. Kleiner K (2009) Peak energy: promise or peril? Nat Rep Clim Change 3:31–33
6. McGlade C, Ekins P (2015) The geographical distribution of fossil fuels unused when limiting global warming to 2 °C. Nature 517(7533):187–190
7. Davis CB, Chmieliauskas A, Dijkema GPJ, Nikolic I (2014) Enipedia. 2014. Delft (NL): Energy & industry group, faculty of technology, policy and management, TU Delft. http://enipedia.tudelft.nl
8. http://www.engineeringnews.co.za/article/report-forecasts-dominant-role-for-gas-in-africas-power-mix-2015-02-27/rep_id:3182
9. http://www.petroleumagencysa.com/index.php/home-14/shale-gas
10. EIA (Energy Information Administration) (2013) Technically recoverable shale oil and shale gas resources: an assessment of 137 shale formations in 41 countries outside the United States report. http://www.eia.gov/analysis/studies/worldshalegas/pdf/fullreport.pdf
11. Pierrot M (2014) The wind power database. http://www.thewindpower.net
12. Creamer T (2014) Eskom weighs gas options as diesel costs double to R10.5bn. http://tinyurl.com/nkhu5ah
13. South African Department of Energy (DoE) (2013) Integrated resource plan for electricity (IRP) 2010–2030. Update Report 2013. http://www.doeirp.co.za/content/IRP2010_updatea.pdf
14. South African Department of Energy, INEP (Integrated National Electrification Programme) (2014) http://www.energy.gov.za/files/policies/p_electricity.html
15. South African Centre for Carbon Capture & Storage (2014) CCS Roadmap. http://www.sacccs.org.za/roadmap/
16. South African Department of Energy (DoE) (2013) Solar-Power. http://www.energy.gov.za/files/esources/renewables/r_solar.html
17. REN21 (2014) Renewables global status report. http://www.ren21.net/gsr
18. Pfenninger S, Gauché P, Lilliestam J, Damerau K et al (2014) Potential for concentrating solar power to provide baseload and dispatchable power. Nat Clim Change 4(8):689–692

19. Gauché P, Backström TW, Brent AC (2013) A concentrating solar power value proposition for South Africa. J Energ Southern Africa 24(1):67–76
20. IRENA (2015) Global Atlas for renewable energy. http://irena.masdar.ac.ae/
21. Staffell I, Green R (2014) How does wind farm performance decline with age? Renew Energ 66:775–786
22. Staffell I, Green R (2015) Is there still merit in the merit order stack? The impact of dynamic constraints on optimal plant mix. IEEE Trans Power Syst. http://dx.doi.org/10.1109/TP WRS.2015.2407613 (in press)
23. Adamson KA (2015) The fuel cell and hydrogen annual review, 2015. 4th Energy Wave. http://www.4thenergywave.co.uk/annual-review/
24. Pollet BG, Pasupathi S, Swart G, Mouton K, Lototskyy M, Bujlo P, Ji S, Bladergroen BJ, Linkov V (2014) Hydrogen South Africa (HySA) systems competence centre: mission, objectives, technological achievements and breakthroughs. Int J Hydrogen Energy 39:3577–3596

Chapter 7
Conclusions and Policy Implications

Abstract This Chapter discusses the policy implications on energy production and usage. It also shows that RSA will have to invest more in its power sector in order to fight the chronic shortage of generating capacity. Although this will cost in the short run, it is shown to save money in the long run in terms of fuel expenditure and improved manufacturing productivity and confidence.

Keywords Policy implications · Manufacturing · Productivity

RSA (and Africa) is currently suffering from a severe Energy crisis which greatly impacts its manufacturing sector, the competitiveness of its economies and capabilities mainly due to (i) capacity constraints and (ii) continually rising energy prices. The increase of energy prices for example for electricity is entirely related to historic under-investment in capacity. The installed fleet of power stations is insufficient to meet the current demand for electricity in a cost-efficient manner, so plants with high variable costs originally designed for peaking are now being used extensively in RSA. The only route to alleviating this capacity shortage is to fund the construction of new and refurbished (very old) power generation facilities, which adds further to the price of electricity. In most parts of West and East Africa, backup power systems (mainly diesel-powered generators) are used by industry and manufacturing companies as their main energy source.

Another alarming challenge in Africa is the use of very expensive fuels to produce electricity, as sadly a number of African countries do not have access to competitively priced fuels such as oil, gas and coal, while Africa also has (unlike the Western World) under-developed intra-regional transport and logistics infrastructures. Consequently this has a detrimental effect on African manufacturing from seriously competing with Asian (mainly China) and the developed world. In order to do so, Africa needs to ensure a stable supply of affordable electricity and tri-ple existing power generation, transmission and distribution infrastructure over the next 30 years.

RSA is a special case in this regard, as it has bountiful supplies of coal and the infrastructure to extract and distribute it. What it lacks is sufficient generating capacity convert this coal into the growing quantity of electricity that the country's economy needs—again due to a prolonged period of underinvestment.

To be *on-par* with the BRIC countries, to "catch up" and to be competitive, Africa needs to adopt a different strategy i.e. to be an independent power producer and to produce power locally in order to sustain its long-term growth ambitions and success. This is due to the fact that investing in power generation infrastruc-ture in the short-term will add significantly to manufacturing capital costs and thus impact on competitiveness. However, in the long-term it does make sense if and only if the incurred cost can be offset by lowering electricity prices, ensuring sta-ble and secure productivity of the current unreliable power supply by most of the African region's state-owned power utilities (e.g. *Eskom* in RSA).

Furthermore, investment in Renewable Energy sources, such as wind and solar farms, is a strategy that (South) African manufacturing companies need to explore over the next 30 years. In countries such as RSA, the set-up of renewable power producing facilities also should come with tax incentives.

Regional Integration for Africa co-operation and development is thus very important and crucial for the realization of the Programme for Infrastructure Development for Africa (PIDA). In the short term the ANC has announced that it will ensure that the following are undertaken:

- Procure up to 20,000 MW of renewable energy, also increase hydro-power imports,
- Revise the National Electrification Plan (NEP), which is aimed at bringing power to all South Africans, with the aim of 97 % electrification by 2025,
- Review the potential for nuclear,
- As long as environmental concerns can be alleviated, fast track development of on-shore (2017) and off-shore (2015) natural gas reserves,
- Update the IRP.

The implications of these, and the Mineral and Petroleum Resources Development Act Methods are that the energy sector in RSA will continue to be in a period of revolution until at least 2020. It is forecasted that by this point RSA will: (a) be home to a home grown renewable energy and fuel cell industries, (b) start to seriously adopt energy storage technologies, (c) have an increasing percentage of energy coming from natural gas, (d) have a decreasing reliance on coal and (e) meet its carbon emission targets (set at COP17 in Copenhagen).

When these are combined with a stronger localised energy industry, and the decreasing dominance of *Eskom*, this could provide a more balanced, and secure energy future for RSA, enabling more robust economic growth.

The challenge in RSA is balancing energy supply and demand and climate change objectives with the very pressing and alarming socio-economic issues, and promoting low carbon growth whilst tackling unemployment, inequality and poverty.

For more than a decade RSA's energy sector has suffered from chronic under-investment in new electricity generation capacity. This is now the root of the two crises facing the economy: rapidly escalating power prices and the threat of supply shortages; both of which harm manufacturing productivity and the economic outlook for the country.

The failure to build sufficient capacity can be traced to two broad causes:

1. Failure of strategy—*Eskom* and the government did not predict the rate at which demand for electricity would grow, or chose to ignore warnings about the impeding capacity shortage. These criticisms have been widely levelled at both parties since the 1990s.
2. Failure in the bureaucratic process—requests from *Eskom* to the government for funds to build new capacity were denied, thus preventing capacity from being built. Attempts to privatise *Eskom* resulted in it being banned from investing in new capacity from the late 1990s until 2004.

Two routes exist to alleviating the crisis: build more dispatchable generating capacity and reduce the demand for electricity. Although it is evident that significant Energy initiatives and Programmes have been introduced in RSA, these are yet to be fully implemented, and there is still a vast gap between the planning and announcement of these efforts and actual implementation. *Eskom* is attempting to go down this first route by initiating projects to build large coal-fired stations: Medupi and Kusile. However, there have been severe delays and cost overruns in their new projects, echoing the experiences in Europe of building a new generation of nuclear reactors.

The REIPPPP is the government's flagship policy to incentivise investment in new renewables capacity. This will have positive outcomes for fuel costs and carbon emissions from the electricity sector, but for the most part it will not help alleviate the capacity shortage, as wind and PV capacity cannot be relied upon to provide power during peak periods (typically winter evenings). In this sense, renewables displace energy, not power. Other renewables, particularly CSP (provided it has sufficient thermal storage), biomass and hydro are dispatchable to an extent, and so will be able to contribute to both energy and peak power requirements. However, these technologies have only made up 11 % of the capacity allocated in bid Windows 1 to 3.

The government's move to allow IPPs into the market is a very positive step. Provided that these are able to secure long term contracts for the purchase of power, building new capacity should be a relatively low-risk investment given the chronic need for power at present. Finance should therefore be forthcoming, allowing IPPs to build new capacity that will be competitive against *Eskom*'s. This will improve diversity in RSA's supply chain of electricity production, moving away from sole reliance on *Eskom*'s ability to finance and construct power stations. It may also improve the diversity in technology choices, both with a move towards gas-fired generation, and also balancing out the large monolithic developments such as Medupi and Kusile with smaller (and quicker to build) power stations.

The other option is to look at effective ways to promote electricity conservation during peak times. In addition to *Eskom*'s efforts to raise awareness through TV advertisements, options exist such as differential tariffs—with a peak rate enforced perhaps just during winter daytime and evenings, and a base rate for all other periods. Electricity storage (e.g. battery and hydrogen) is a technological solution that is rapidly gaining traction in Germany due to the issues caused by uncontrollable PV installations. Residential or utility-scale storage could be used effectively to shift demand from the evening period to later at night, allowing peak demand to grow without having to expand generation and transmission capacity.

In the longer term, significant shifts are being planned which could help to diversify and strengthen the RSA energy sector. Natural gas has the potential to become a significant energy vector in the near future, and could be the springboard for the low carbon economy. Reserves are currently estimated to be of a similar magnitude to those of coal, the equivalent of 13.2 GT of coal. Whether these reserves come into play depends partly upon geology and technical constraints, but more importantly on the willingness of industry to invest in this area. The RSA government is hoping for significantly more state involvement than has been seen in the US shale gas revolution, with 20 % state ownership of all projects. The government's insistence on the right to majority ownership may end up stifling investment, and thus the country's development of a lower-cost and lower-carbon energy source.

Extremely large new build projects exist for both nuclear, with six new power stations announced at a cost of US$40–90 billion, and plans for the Grand Inga Damn in the DRC at a cost of US$80 billion. These have the potential to bring a large amount of electricity generation capacity online in the coming decade. For example, the 40 GW hydro project is around half the size of the entire Southern African Power Pool (covering 12 countries), and so would have a substantial impact on reducing electricity prices in the long-run, provided that sufficient transmission capacity was built to distribute the electricity to consumers across the region. This could be seen as a long-term solution to three problems in the region: a shortage of capacity, power price escalation linked to global fuel prices, and tackling GHG emissions. Both the nuclear and hydro projects are capital intensive, and so will rely on the ability of governments to raise enormous amounts of capital at a time when national debts are high and the prospects for future growth look uncertain.

It is unavoidable that RSA will have to invest more in its power sector in order to combat the chronic shortage of generating capacity. Although this will cost in the short run, it is shown to save money in the long run in terms of fuel expenditure and improved manufacturing productivity and confidence.

Appendix A: Detailed Power Sector Data

Table A.1 lists the power stations in operation in South Africa as of 2012, giving the installed capacity by type, and the expected date of decommissioning.

Table A.1 Current power plants in South Africa, 2012

Type of station	Name of station	Nominal capacity (MW)	Current proposed decommissioning schedule
Coal—baseload	Arnot	2,232	2025–2029
Coal—baseload	Duvha	3,450	2030–2034
Coal—baseload	Hendrina	1,865	2021–2027
Coal—baseload	Kelvin	600	
Coal—baseload	Kendal	3,840	2027, 2038–2043
Coal—baseload	Kriel	2,850	2026–2029
Coal—baseload	Lethabo	3,558	2035–2040
Coal—baseload	Majuba	3,843	2046–050
Coal—baseload	Matimba	3,690	2023, 2037–2041
Coal—baseload	Matimba B	3,990	
Coal—baseload	Matla	3,450	2029–2033
Coal—baseload	Pretoria West	180	
Coal—baseload	Rooiwal	300	
Coal—baseload	Tutuka	3,510	2024, 2035–2040
Coal—baseload	Medupi	4,788	
Coal—baseload	Kusile	4,800	
Nuclear-baseload	Koeberg	1,940	
Coal (return to service)—baseload	Camden	1,510	2020–2023
Coal (return to service)—baseload	Grootvlei	1,200	2025–2028
Coal (return to service)—baseload	Komati	940	2024–2028

(continued)

B.G. Pollet et al., *The Energy Landscape in the Republic of South Africa*, SpringerBriefs in Energy, DOI 10.1007/978-3-319-25510-1

Table A.1 (continued)

Type of station	Name of station	Nominal capacity (MW)	Current proposed decommissioning schedule
Hydroelectric—peaker	Gariep	360	
Hydroelectric	Steebras	180	
Hydroelectric—peaker	Vanderkloof	240	
Pumped Storage—peaker	Ingula	1,332	
Pumped Storage—peaker	Drakensberg	1,000	
Pumped Storage—peaker	Palmiet	400	
Natural Gas Turbine—peaker	Ankerlig	1,338	
Gas	Avon	670	
Gas	Dedisa	335	
Natural gas turbine—peaker	Gourikwa	746	
Natural gas turbine—peaker	Acacia	171	
Natural gas turbine—peaker	Port Rex	171	
Renewables—wind (under construction)	Amakhala Emoyeni	134.4	
Renewables—wind (under construction)	Coega	43.2	
Renewables—wind (under construction)	Cookhouse	138.6	
Renewables—wind (under construction)	Dorper	100	
Renewables—wind	Eastern Cape	80	
Renewables—wind (under construction)	Eastern Cape #2	27	
renewables—wind	Grassridge	61.5	
Renewables—wind	Hopefield	66.6	
Renewables—wind	Klipheuwel	30.1	
Renewables—wind	Klipheuwel	3	
Renewables—wind	Nobelsfontaine	73.8	
Renewables—wind	Sere	100	
Renewables—wind	Van Stadens	27	
CSP		100	
CSP (under construction)	Bokpoort	54.5	
CSP (under construction)	KaXu Solar One	100	

(continued)

Table A.1 (continued)

Type of station	Name of station	Nominal capacity (MW)	Current proposed decommissioning schedule
CSP (under construction)	Khi Solar One	50	
CSP (under construction)	Xina Solar One	100	
Hydroelectric	First Falls	6	
Hydroelectric	Second Falls	11	
Hydroelectric	Colley Wobbles	42	
Hydroelectric	Ncora	2	
	Total MWs	45,710	

Source Eskom, South African department of energy and international energy agency

Table A.2 lists the renewable power projects that have been awarded under the REIPPPP bid windows, grouped by type.

Table A.2 South African REIPPPP Bid Windows 1, 2 and 3 projects by renewable energy technology. *Source* Eberhard [8]

	Project name	Capacity awarded (MW)
Wind	1. Nobelsfontein Phase 1	75
	2. Dorper Wind Farm	97.5
	3. Dassieklip Wind Energy Facility	27
	4. Metrowind Van Stadens Wind Farm Onshore	27
	5. Kouga Red Cap Wind Farm—Oyster Bay	80
	6. Jeffreys Bay Onshore Wind	138
	7. Hopefield Wind Farm Onshore Wind	65.4
	8. Cookhouse Wind Farm Onshore Wind	138.6
	9. Gouda Wind Project	135.5
	10. Amakhala Wind Project	133.7
	11. Tsitsikamma Community Wind Farm	94.8
	12. Wind Farm West Coast 1	90.8
	13. Waainek Wind Power	23.3
	14. Grassridge Onshore Wind Project	59.8
	15. Chaba Wind Power	21
	16. Longyuan Mulilo Green Energy De Aar 2 North Wind	139
	17. Longyuan Mulilo De Aar Maanhaarberg Wind Energy	96.5
	18. Nojoli Wind Farm	96.5
	19. Loeriesfontein 2	138.2
	20. Noupoort	79.1
	21. Khobab Wind	137.7
	22. Red Cap—Gibson Bay	110

(continued)

Table A.2 (continued)

	Project name	Capacity awarded (MW)
Photovoltaic (PV)	23. Letsatsi Solar Photovoltaic Park Photovoltaic	64
	24. Lesedi Solar Photovoltaic Park Photovoltaic	64
	25. Witkop Solar Park	30
	26. Touwsrivier Solar Park	36
	27. Soutpan Solar Park	28
	28. Mulilo Solar PV De Aar	10
	29. Mulilo Solar PV Prieska	20
	30. Konkoonsies Solar Energy Facility	9.7
	31. RustMo1 Solar Farm	6.9
	32. Kalkbult	72.5
	33. Aries Solar Energy Facility	9.7
	34. Slimsun Swartland Solar Park	5
	35. Mainstream Renewable Power De Aar PV	45.6
	36. Greefspan PV Power Plant	9.9
	37. Kathu Solar Plant	75
	38. Solar Capital De Aar	75
	39. Mainstream Renewable PowerDroogfontein	45.6
	40. Herbert PV Power Plant	20
	41. Solar Capital De Aar 3	75
	42. Sishen Solar Facility	74
	43. Aurora-Rietvlei Solar Power	9
	44. Vredendal Solar Park	8.8
	45. Linde	36.8
	46. Dreunberg	69.6
	47. Jasper Power Company	75
	48. Boshoff Solar Park	60
	49. Upington Airport	8.9
	50. Adams Solar PV 2	75
	51. Electra Capital (Pty) Ltd	75
	52. Mulilo Sonnedix Prieska PV	75
	53. Mulilo Prieska PV	75
	54. Tom Burke Solar Park	60
	55. Pulida Solar Park	75
CSP[a]	56. Kaxu Solar One	100
	57. Khi Solar One	50
	58. Bokpoort CSP project	50
	59. Ilanga CSP 1 / Karoshoek Solar One	100
	60. !XiNa Solar One	100
Other	61. Stortemelk Hydro Power Plant	4.4
	62. Neusberg Hydro Electrical Project	10
	63. Mkuze Biomass	16.5
	64. Joburg Landfill Gas to Electricity	18

[a]Concentrated solar power

Printed in the United States
By Bookmasters